ISBN 978-1-5282-9777-6
PIBN 10966997

1 MONTH OF FREE READING

at

www.ForgottenBooks.com

By purchasing this book you are eligible for one month membership to ForgottenBooks.com, giving you unlimited access to our entire collection of over 1,000,000 titles via our web site and mobile apps.

To claim your free month visit:
www.forgottenbooks.com/free966997

English
Français
Deutsche
Italiano
Español
Português

www.forgottenbooks.com

Mythology Photography **Fiction**
Fishing Christianity **Art** Cooking
Essays Buddhism Freemasonry
Medicine **Biology** Music **Ancient**
Egypt Evolution Carpentry Physics
Dance Geology **Mathematics** Fitness
Shakespeare **Folklore** Yoga Marketing
Confidence Immortality Biographies
Poetry **Psychology** Witchcraft
Electronics Chemistry History **Law**
Accounting **Philosophy** Anthropology
Alchemy Drama Quantum Mechanics
Atheism Sexual Health **Ancient History**
Entrepreneurship Languages Sport
Paleontology Needlework Islam
Metaphysics Investment Archaeology
Parenting Statistics Criminology
Motivational

Technical Paper 16

DEPARTMENT OF THE INTERIOR

BUREAU OF MINES

JOSEPH A. HOLMES, Director

DETERIORATION AND SPONTANEOUS HEATING OF COAL IN STORAGE

A PRELIMINARY REPORT

BY

HORACE C. PORTER

AND

F. K. OVITZ

WASHINGTON

GOVERNMENT PRINTING OFFICE

1912

CONTENTS.

First edition.   April, 1912.

DETERIORATION AND SPONTANEOUS HEATING OF COAL IN STORAGE.

By Horace C. Porter and F. K. Ovitz.

GENERAL STATEMENT.

Various departments and establishments of the United States Government, particularly the Navy and the Panama Railroad Co., are heavy consumers of coal and are compelled to store large quantities under climatic conditions likely to cause deterioration. Spontaneous heating frequently occurs in heaps or bins and the resulting fires not only waste coal but may endanger life. Hence the Government has a direct interest in any investigations concerning the deterioration of coal in storage, and the investigation by the Bureau of Mines was undertaken primarily in order to increase economy, efficiency, and safety in the use of coal by the Government.

Coal-storage problems have grown in importance with the increasing consumption of coal and the consequent greater need of guarding against interruptions of supply by strikes or transportation difficulties. The naval coaling stations, the Panama Railroad Co., the commercial coal-distributing companies on the Great Lakes, the large coke, gas, or power plants at a distance from the coal fields, and many of the railroads, particularly those in the West, keep 50,000 to 500,000 tons in storage a large part of the time.

Owing to the large number of requests that are being received by the Bureau of Mines for exact information on the deterioration and spontaneous heating of coal, this preliminary paper is published as a brief summary of the results thus far obtained in the bureau's investigation. A complete account is being prepared for publication as a bulletin of this bureau. The bulletin will present in detail the results of a large number of analyses and heating-value determinations, descriptions of methods of conducting tests, and a brief résumé of the important previous investigations on the subject. It will contain also a tabulated summary of information on cases of spontaneous combustion of coal. This information was obtained through a letter of inquiry sent by the bureau in January, 1911, to over 2,000 of the largest coal consumers in the United States.

DETERIORATION OF COAL.

VARIETY OF OPINIONS.

Not many years ago coal was commonly regarded as an extremely unstable material, subject to serious alteration and losses on exposure to the elements. E. C. Pechin, speaking before the American Institute of Mining Engineers, in 1872, said: [a]

> Fuel suffers materially by * * * storage; especially with bituminous and semi-bituminous coals is the loss heavy, an exposure of only two weeks causing a loss of carbon to the extent of 10 to 25 per cent.

Similar views have been held more recently. For example, in a paper before the United States Naval Institute, in 1906, [b] one finds the statement:

> The pressure of the weight of coal causes gases to be evolved; * * * these gases constitute the chief and only value of the coal in that they furnish the heat units. It is claimed that if a ton of fine bituminous coal be spread out on a concrete pavement * * * in the open air in this climate [Key West, Fla.] for one year, it will lose all its calorific properties. The gases are simply free to escape, and when the coal has lost all its gas it will have lost all its heat units and be simply coke.

The author's meaning, no doubt, is that the heat units in the volatile part of the coal are all lost by storage in the manner described.

In 1907 a German gas-works engineer claimed [c] to have found that moist fine coal sustained an average heating value loss of 1.7 per cent per week by the escape of gas. The 1889 edition of Groves and Thorpe's "Chemical Technology of Fuels" says: [d] "In some places coal is known to lose 50 per cent of its heating value in six months."

Other statements like these are to be found in recent literature, but probably the great majority of chemists and engineers to-day hold no such exaggerated ideas on the deterioration of coal in storage. There is, however, a well-defined suspicion in the minds of many that the loss of volatile matter by coal and the deterioration by oxidation are large enough to be of industrial importance, and the object of the investigations described in this paper was to determine accurately the extent of the deterioration in different types of coal.

INVESTIGATIONS OF THE BUREAU OF MINES.

A study was made in the laboratory of the loss of volatile matter from crushed coal during storage. The coal tested was in 20-pound samples which represented a variety of types from widely separated fields. Each sample had been broken to about half-inch size in the mine and as quickly as possible placed in a large glass bottle for shipment to the laboratory. At the laboratory the accumulated gas

---

a Trans. Am. Inst. Min. Eng., vol. 1, p. 286.

b Proc. U. S. Naval Inst., vol. 32, pp. 545–554.

c Hannack, F., Am. Gas Light Jour., vol. 86, p. 843. Abstracted from Stahl und Eisen.

d Vol. I, p. 83.

was withdrawn and the volatile products were permitted to escape freely and continuously at atmospheric pressure and temperature. The results of these experiments have been published in a technical paper of the Bureau of Mines,[a] and therefore will not be given here in detail. Suffice it to say that although several coals evolve methane in large volumes, especially in the early period after mining, the coals tested lost in one year from this cause only 0.16 per cent at most of their calorific value.

It seems, therefore, that the loss caused by escape of volatile matter from coal has been greatly overestimated.

At the instance of the Navy Department, which purchases two· or three million dollars' worth of coal annually, and stores large quantities in warm climates for long periods of time, more elaborate tests were undertaken to determine the total loss possible in high-grade coal by weathering. The extent of the saving to be accomplished by water submergence as compared with open-air storage was to be settled, as was also the question whether salt water possessed any peculiar advantage or disadvantage over fresh water for this purpose. As an instance of the claims made for the advantage of salt-water submergence there may be cited an article by an English railway and dock superintendent,[b] who reports that coal accidentally sunk in the salt mud of the English Channel actually had its calorific value improved 1.8 per cent by 10 years' submergence, "which offers a very pretty question to chemists to say why." He claims that salt preserves the virtues of coal as it does those of many other things, and that "if coal is given a strong dose of coarse salt and water 12 hours before using its calorific value is greatly improved."

OUTLINE OF TESTS.

In brief outline, the tests by the bureau were carried out as follows: Four kinds of coal were chosen—New River, W. Va., on account of its large use by the Navy; Pocahontas, Va., from its wide use as a steaming and coking coal in the Eastern States, and from its being the principal fuel used in the Panama Canal work; Pittsburg coal, as a rich coking and gas coal, and Sheridan, Wyo., subbituminous coal or "black lignite," as a type much used in the West. In the tests with the New River coal, 50-pound portions were made up out of one large lot, which had been crushed to quarter-inch size and well mixed, and these portions, confined in perforated wooden boxes were submerged under sea water at three navy yards that differed widely from one another in climatic conditions. Portions of 300 pounds each from the same

---

[a] Porter, H. C. and Ovitz, F. K., The escape of gas from coal. Technical Paper 2, Bureau of Mines, 1911, 14 pp.

[b] Macaulay, J., Engineer (London), vol. 96, pp. 238, 415.

original lot were exposed to the open air, both outdoors and indoors, at the same navy yards.

Test of the Pocahontas coal was made at only one point, the Isthmus of Panama, where run-of-mine coal was placed in a 120-ton pile exposed to the weather. Pittsburg coal was stored as run-of-mine at Ann Arbor, Mich., both in open 5-ton bins outdoors and in 300-pound barrels submerged in fresh water. The Wyoming subbituminous coal was stored at Sheridan, Wyo., both as run-of-mine and slack, in outdoor bins that held 3 to 6 tons each.

Duplicate samples were taken at each sampling, and the sampling of every lot, except the outdoor pile at Panama and the 300-pound open-air piles at the navy yards, was done by rehandling all the coal. In the excepted cases it was not thought fair to disturb the entire lot. Therefore at Panama a vertical section of the pile, weighing 10 tons was removed each time; eight gross samples were taken from this 10-ton portion, and each of the samples was quartered down. From the outdoor piles at the navy yards a number of small portions were taken from well-distributed points in each pile, mixed, and quartered down. Small lots and a fine state of division were purposely adopted with the New River coal, so as to make the tests of maximum severity.

A fact the authors wish to emphasize is that no test was made for any physical deterioration of the coal, such as an increase of friability, or a slacking of lumps. Consequently the results do not show how the relative availability of heat units may have been affected by physical changes; but they do show how much the calorific value, as determined in the laboratory, was affected by storage of the coal under the several conditions.

RESULTS OF TESTS.

Table 1 gives a condensed summary compiled from results of a large number of tests and determinations. The moisture, ash, and sulphur contents and the calorific value of each sample were determined. A Mahler bomb calorimeter and a carefully calibrated Beckmann thermometer were used in the calorimetric work, and except for the Sheridan, Wyo., tests, all this work was done by the same man (one of the authors), and with the same calorimeter. To eliminate the effect of incidental impurities that were not concerned in the deterioration of the actual coal substance, the calorific values are calculated to a moisture, ash, and sulphur free basis, and the ash as determined is corrected for errors due to oxidation of pyrites and dehydration of shale or clay.

TABLE 1.—*Calorific value (in gram calories) of the coal substance (moisture, ash, and sulphur free coal) before and after storage.*

Kind of coal.	Storage conditions.	Place.	Calorific value.				Per centage of loss in 2 years.
			As stored.	After 3 months.	After 1 year.	After 2 years.	
New River, W.Va.a	¼-inch coal, under sea water.	Portsmouth, N. H..	8,761	....	8,759	8,730	b 0.39
		Norfolk, Va..........	8,754	....	8,763	8,737	b .40
		Key West, Fla......	8,747	....	8,732	8,770	0
	¼-inch coal, under fresh water.	Pittsburgh, Pa......	8,752	8,762	8,749	8,760	0
	¼-inch coal, exposed to air indoors.	Portsmouth, N. H..	8,779	8,738	8,742	8,709	.80
		Norfolk, Va..........	8,751	8,742	8,736	8,717	.39
		Key West, Fla......	8,754	8,728	8,676	....	....
		Pittsburgh, Pa......	8,769	8,736	8,719	8,720	.56
	R. o. m. coal, exposed to air indoors.	Portsmouth, N. H..	8,754	8,758	8,748	8,734	.23
		Norfolk, Va..........	8,743	8,764	8,725	8,725	.21
		Key West, Fla......	8,745	8,733	8,710	....	....
		Pittsburgh, Pa......	8,753	8,758	8,740	8,743	.11
	¼-inch coal, exposed to air outdoors, uncovered.	Portsmouth, N. H..	8,741	8,730	8,701	8,674	.77
		Norfolk, Va..........	8,726	8,709	8,666	8,625	1.16
		Key West, Fla......	8,739	8,721	8,680	8,592	1.85
		Pittsburgh, Pa......	8,763	8,726	8,700	8,665	1.12
	R. o. m. coal, exposed to air outdoors, uncovered.	Portsmouth, N. H..	8,775	8,766	8,760	8,722	.60
		Norfolk, Va.	8,743	8,745	8,720	8,695	.55
		Key West, Fla......	8,752	8,708	8,722	8,632	1.29
		Pittsburgh, Pa......	8,752	8,740	8,716	8,696	.64
Pocahontas, Va.....	120-ton pile, r. o. m. coal, outdoors.	Panama.............	8,794	8,787	8,762	....	....
Pittsburg bed......	4 tons r. o. m. coal in open bin, outdoors.	Ann Arbor, Mich....	8,541	c 8,539	8,526	....	....
			8,541	d 8,553	8,528	....	....
Sheridan, Wyo., subbituminous. e	R.o.m., open bin 5 ft. deep.	Sheridan, Wyo......	7,370	7,174	....	7,135	f 3.19
	R.o.m.,closed bin 5 ft. deep.	.....do	7,370	7,202	....	7,094	f 3.75
	R.o. m.,closed bin 15 ft. deep.	.....do	7,370	7,303	....	6,982	f 5.26
	Slack, closed bin 15 ft. deep.	.....do	7,355	7,166	....	6,990	f 4.96

a Heating value of representative mine sample of coal (ash, moisture, and sulphur free), 8,768 calories.
b On basis of original heating value of same test portion.
c After six months' storage, entire lot in bin.
d After six months' storage, upper 6 inches at surface of bin.
e Each bin contained 5 to 10 tons.
f After two years and three-quarters.

The results show in the case of the New River coal a loss of less than 1 per cent of calorific value in one year by weathering in the open. In two years the greatest loss was at Key West, 1.85 per cent. There was practically no loss at all in the submerged samples in one year, fresh or salt water serving equally well to "preserve the virtues" of the coal. There was almost no slacking of lumps in the run-of-mine samples. In all tests the crushed coal deteriorated more rapidly than the run-of-mine.

The Pocahontas run-of-mine coal in a 120-ton pile on the Isthmus of Panama lost during one year's outdoor weathering less than 0.4 per cent of its heating value, and showed little slacking of lumps.

Gas coal during one year's outdoor exposure suffered practically no loss of calorific value, measurable by the calorimetric method used, not even in the coal forming the top 6-inch layer in the bins.

The Wyoming coal in one of the bins deteriorated in heat value 5.3 per cent during storage for two and three-fourths years, and more than 2.5 per cent in the first three months.   There was bad slacking and crumbling of the lumps on the surface of the piles, but even where this surface was fully exposed to the weather the slacking did not penetrate more than 12 to 18 inches in the 2¾-year period.

No outdoor weathering tests of coal of the Illinois type, have been made by the Bureau of Mines, but tests of such coal have been reported by Prof. S. W. Parr, of the University of Illinois[a] and by A. Bement, of Chicago,[b] both of whom find a calorific loss of 1 to 3 per cent in a year by weathering.   Bement reports a slacking of lumps of over 80 per cent in one test and of about 12 per cent in another test of small samples.   It is possible that with the Illinois as with the Wyoming coal, the slacking in a large pile would not penetrate far from the surface.

Storage under water unquestionably preserves the heating value and the physical strength of coal.   But such storage practically makes necessary the firing of wet coal, and consequently the evaporation in the furnace of added moisture varying in amount from 1 to 15 per cent according to the kind of coal.   This factor is an important drawback to the underwater storage of coals like the Illinois and Wyoming types which mechanically retain 5 to 15 per cent of water after draining.   In the case of high-grade eastern coals, however, if firemen are permitted, as is ordinarily the case, to wet down their coal before firing, in order, as some say, "to make a hotter fire," the addition during storage of the 2 or 3 per cent of moisture which these coals retain becomes of little consequence.   Submerged storage is an absolute preventive of spontaneous combustion, and on that account alone it may be justified when the coal is particularly dangerous to store and when large quantities are to be stored; but unless the storage period is to be longer than one year, there seems to be no ground for storing any coal under water merely for the sake of the saving in calorific value to be obtained by the avoidance of weathering.

SPONTANEOUS COMBUSTION.

UNDERLYING CAUSES.

Losses of value from spontaneous heating are a much more serious matter than the deterioration of coal at ordinary temperatures.   Oxidation proceeds more rapidly as the temperature rises.   The oxidation, beginning at ordinary temperatures, attacking the surfaces of particles, and developing heat, is probably, in some degree, an absorption of oxygen by the unsaturated chemical compounds in the coal substance.

a Bulletin 38, Univ. of Ill., Eng. Expt. Sta., 1909.
b Chemical Engineer, vol. 12, July, 1910, p. 9.

In a small pile of coal this slowly developed heat can be readily dissipated by convection and radiation and very little rise in temperature results. If the dissipation of the heat is restricted, however, as in a large pile densely packed, the temperature within the pile rises continuously. The rate of oxidation of the coal platted against the temperature makes a curve that rises with great rapidity. When the storage conditions are such as to allow warming of the coal to a temperature of about 100° C., the rate of oxidation becomes so great that the heat developed in a given time ordinarily exceeds the heat dissipated, and the temperature rises until, if the air supply is adequate, the coal takes fire. Evidently, therefore, it is important to guard against even moderate heating of the coal, either spontaneous or from an external source. Increased loss of volatile matter and of heating value occurs with a moderate rise of temperature, even though the ignition point is not reached.

Spontaneous combustion is brought about by slow oxidation in an air supply sufficient to support oxidation but insufficient to carry away all the heat formed. The area of surface exposed to oxidation by a given mass of any one coal determines largely the amount of oxidation that takes place in the mass; it depends on the size of the particles and increases rapidly as the fineness approaches that of dust. Dust is therefore dangerous, particularly if it is mixed with lump coal of such a size that the interstices permit the flow of a moderate amount of air to the interior. Coals differ widely in friability, that is, in the proportion of dust that is produced under like conditions. In comparative tests samples of Pocahontas, New River, and Cambria County, Pa., coals produced nearly twice as much dust (coal through a ⅛-inch screen) as coal mined from the Pittsburg bed in Allegheny County, Pa. This variation in friability is a factor in affecting the liability to spontaneous heating.

Ideal conditions for such heating are offered by a mixture of lump and fine coal, such as run-of-mine with a large percentage of dust, piled so that a small supply of air is admitted to the interior of· the pile.

EFFECT OF VOLATILE MATTER.

High volatile matter does not of itself increase the liability of coal to spontaneous heating. A letter of inquiry sent by the bureau to more than 2,000 large coal consumers in the United States brought 1,200 replies. Of the replies, 260 reported instances of spontaneous combustion and 220 of the 260 gave the name of the coal. The 220 instances were distributed as follows: Ninety-five were in semibituminous low-volatile coals of the Appalachian region, 70 in higher-volatile coals of the same region, and 55 in western and middle western coals. This result shows at least no falling behind of the "smokeless" type of coal in furnishing instances of spontaneous combustion, and no cause for placing especial confidence in this type of coal for safety in storage.

A serious fire in a cinder filling under a manufacturing plant in Pittsburgh, Pa., was recently investigated by the Bureau of Mines. All the evidence indicated that the cause was spontaneous combustion induced by heat transmitted from an external source—a gas-fired furnace in contact with the filling. The cinders contained 40 per cent of carbon. A similar fire occurred in April, 1910, in a cinder filling under a smelting plant on Staten Island. In this case the cinders contained 33 per cent of carbon. Damage amounting to $20,000 was done. The cause was not definitely determined, but from the reports of the insurance adjusters spontaneous heating appears to be the most plausible explanation. In these two cases the volatile matter in the material was probably not a factor.

In the large stock piles at Panama, Appalachian coals, with 17 to 21 per cent volatile matter, give a great deal of trouble from spontaneous fires. Moreover, several large works report that their low-volatile coals are more troublesome in respect to spontaneous fires than their high-volatile gas coals. The high-volatile coals of the West are, to be sure, usually very liable to spontaneous heating, but owe this property to the chemical nature of their constituent substances rather than to the amount of volatile matter that they contain. Strange as it may seem, the oxygen content of coal appears to bear a direct relation to the avidity with which coal absorbs oxygen; high oxygen coals absorb oxygen readily, and therefore have a marked tendency to spontaneous combustion.

EFFECT OF MOISTURE AND OF SULPHUR.

The effect of moisture and the effect of sulphur on the spontaneous heating of coal are questions on which there has been a great deal of discussion and much difference of opinion. Very little experimental evidence has been brought to bear on either of the questions, and certainly neither is as yet settled. Richters has shown [a] that in the laboratory dry coal oxidizes more rapidly than moist; but the weight of opinion among practical users of coal is that moisture promotes spontaneous heating. In not one of the many cases of spontaneous combustion observed by the authors, as representatives of the Bureau of Mines, could it be proved that moisture had been a factor. Still the physical effects of moisture on fine coal, such as closer packing together of dust or small pieces, may in many cases aid spontaneous heating.

Sulphur has been shown by the bureau's investigations to have only a minor effect in most instances. On a number of occasions samples of coal from the places in a pile or bin where the heat was greatest have been analyzed, both for the total sulphur and for that in the sulphate or oxidized form. The difference between the two determinations—or, in other words, the unoxidized sulphur—was in no case less than 75 per cent of the average total sulphur in the coal which had not been heated. In other words, not more than one-fourth of the total sulphur had entered into the heat-producing reaction. The possibility remains, of course, that the sulphur-bearing material that oxidized was concentrated in a pocket, possibly as moist, finely divided pyrites, and by its concentration and, perhaps, by its comparatively rapid oxidation sufficient heat was developed in that spot to act as an igniter. Such an explanation, however, is not plausible when judged by the many analyses of coal subjected to spontaneous heating. For example, a Boston company that uses a Nova Scotia coal containing 3 to 4 per cent sulphur has much trouble with spontaneous fires in storage piles. Analyses of a number of samples taken by representatives of the Bureau of Mines from exposed piles in which heating had occurred showed that 90 per cent of the sulphur was still unoxidized.

In certain laboratory experiments, in which air was passed over samples of coal heated to 120° C., enough heat developed to bring the coal almost to the ignition point; the coal was then allowed to cool. Subsequent analysis showed practically no increase in the sulphate or oxidized form of sulphur and practically no reduction of the total sulphur in the coal. Though the results of these experiments are not entirely conclusive, they indicate that sulphur content contributes very little to the spontaneous heating of coal.

IMPORTANCE OF FRESHLY EXPOSED SURFACES.

Freshly mined coal and the fresh surfaces exposed by crushing lumps exhibit a remarkable avidity for oxygen, but after a time the surfaces become coated with oxidized material, "seasoned," as it were, so that the action of the air becomes much less vigorous. In practice, coal that has been stored for six weeks or two months and has even become already somewhat heated, if rehandled and thoroughly cooled by the air, seldom heats spontaneously again.

A large power plant in the New York district crushes its coal to pass a 4-inch screen immediately after unloading from barges and leaves the fines and dust, 50 per cent or more of the total, in the coal to be stored. This freshly crushed coal is elevated to iron hopper-shaped bunkers in which, as they are directly over the boilers, the air temperature is often 100° F. As the coal may hang on the sloping sides of a bunker for three or four months, it is not surprising that there are fires in one or another of the bunkers practically all of the time.

SUGGESTIONS ON STORING COAL.

With full appreciation of the fact that any or all of the following suggested precautions may prove impracticable or unreasonably expensive under certain conditions, they are offered as advisable for safety in storing bituminous coal.

1. Do not pile over 12 feet deep, nor so that any point in the interior of a pile will be over 10 feet from an air-cooled surface.

2. If possible, store only screened lump coal.

3. Keep out dust as much as possible; to this end reduce handling to a minimum.

4. Pile so that lump and fine are distributed as evenly as possible; not, as is often done, allowing lumps to roll down from the peak and form air passages at the bottom of the pile.

5. Rehandle and screen after two months, if practicable.

6. Do not store near external sources of heat, even though the heat transmitted be moderate.

7. Allow six weeks' "seasoning" after mining and before storing.

8. Avoid alternate wetting and drying.

9. Avoid admission of air to interior of pile through interstices around foreign objects, such as timbers or irregular brickwork, or through porous bottoms, such as coarse cinders.

10. Do not try to ventilate by pipes, or more harm may often be done than good.

PUBLICATIONS ON FUEL TESTING.

PUBLICATIONS OF THE BUREAU OF MINES.

The following publications can be obtained free of cost by applying to the Director, Bureau of Mines, Washington, D. C.

BULLETIN 1. The volatile matter of coal, by H. C. Porter and F. K. Ovitz. 1910. 56 pp., 1 pl.

BULLETIN 2. North Dakota lignite as a fuel for power-plant boilers, by D. T. Randall and Henry Kreisinger. 1910. 42 pp., 1 pl.

BULLETIN 3. The coke industry of the United States as related to the foundry, by Richard Moldenke. 1910. 32 pp.

BULLETIN 4. Features of producer-gas power-plant development in Europe, by R. H. Fernald. 1910. 27 pp., 4 pls.

BULLETIN 5. Washing and coking tests of coal at Denver, Colo., by A. W. Belden, G. R. Delamater, J. W. Groves, and K. M. Way. 1910. 62 pp.

BULLETIN 6. Coals available for the manufacture of illuminating gas, by A. H. White and Perry Barker. 1911. 77 pp., 4 pls.

BULLETIN 7. Essential factors in the formation of producer gas, by J. K. Clement, L. H. Adams, and C. N. Haskins. 1911. 51 pp., 1 pl.

BULLETIN 8. The flow of heat through furnace walls, by W. T. Ray and Henry Kreisinger. 1911. 32 pp.

BULLETIN 9. Recent development of the producer-gas power plant in the United States, by R. H. Fernald. 82 pp., 2 pls. Reprint of United States Geological Survey Bulletin 416.

BULLETIN 11. The purchase of coal by the Government under specifications, by G. S. Pope. 80 pp. Reprint of United States Geological Survey Bulletin 428.

BULLETIN 12. Apparatus and methods for the sampling and analysis of furnace gases, by J. C. W. Frazer and E. J. Hoffman. 1911. 22 pp.

BULLETIN 13. Résumé of producer-gas investigations, October 1, 1904, to June 30, 1910, by R. H. Fernald and C. D. Smith. 1911. 393 pp., 12 pls.

BULLETIN 14. Briquetting tests of lignite at Pittsburgh, Pa., 1908–9; with a chapter on sulphite-pitch binder, by C. L. Wright. 1911. 64 pp., 11 pls.

BULLETIN 16. The uses and value of peat for fuel and other purposes, by C. A. Davis. 1911. 214 pp., 1 pl.

BULLETIN 19. Physical and chemical properties of the petroleums of the San Joaquin Valley, Cal., by I. C. Allen and W. A. Jacobs; with a chapter on analyses of natural gas from the southern California oil fields, by G. A. Burrell. 1911. 60 pp., 2 pls.

BULLETIN 21. The significance of drafts in steam-boiler practice, by W. T. Ray and Henry Kreisinger. 62 pp. Reprint of United States Geological Survey Bulletin 367.

BULLETIN 24. Binders for coal briquets, by J. E. Mills. 56 pp. Reprint of United States Geological Survey Bulletin 343.

BULLETIN 27. Tests of coal and briquets as fuel for house-heating boilers, by D. T. Randall. 45 pp., 3 pls. Reprint of United States Geological Survey Bulletin 366.

BULLETIN 28. Experimental work conducted in the laboratory of the United States fuel-testing plant at St. Louis, Mo., January 1, 1905, to July 31, 1906, by N. W. Lord. 49 pp. Reprint of United States Geological Survey Bulletin 323.

BULLETIN 29. The effect of oxygen in coal, by David White. 80 pp., 3 pls. Reprint of United States Geological Survey Bulletin 382.

BULLETIN 30. Briquetting tests at the United States fuel-testing plant, Norfolk, Va., 1907–8, by C. L. Wright. 41 pp., 9 pls. Reprint of United States Geological Survey Bulletin 385.

BULLETIN 31. Incidental problems in gas-producer tests, by R. H. Fernald, C. D. Smith, J. K. Clement, and H. A. Grine. 29 pp. Reprint of United States Geological Survey Bulletin 393.

BULLETIN 32. Commercial deductions from comparisons of gasoline and alcohol tests on internal-combustion engines, by R. M. Strong. 38 pp. Reprint of United States Geological Survey Bulletin 392.

BULLETIN 33. Comparative tests of run-of-mine and briquetted coal on the torpedo boat *Biddle*, by W. T. Ray and Henry Kreisinger. 49 pp. Reprint of United States Geological Survey Bulletin 403.

BULLETIN 34. Tests of run-of-mine and briquetted coal in a locomotive boiler, by W. T. Ray and Henry Kreisinger. 32 pp. Reprint of United States Geological Survey Bulletin 412.

BULLETIN 35. The utilization of fuel in locomotive practice, by W. F. M. Goss. 28 pp. Reprint of United States Geological Survey Bulletin 402.

BULLETIN 37. Comparative tests of run-of-mine and briquetted coal on locomotives, including torpedo-boat tests and some foreign specifications for briquetted fuel, by W. F. M. Goss. 1908. 57 pp., 4 pls. Reprint of United States Geological Survey Bulletin 363.

TECHNICAL PAPER 1. The sampling of coal in the mine, by J. A. Holmes. 1911. 18 pp.

TECHNICAL PAPER 2. The escape of gas from coal, by H. C. Porter and F. K. Ovitz. 1911. 14 pp.

TECHNICAL PAPER 3. Specifications for the purchase of fuel oil by the Government, with directions for sampling oil and natural gas, by I. C. Allen. 1911. 13 pp.

TECHNICAL PAPER 5. The constituents of coal soluble in phenol, by J. C. W. Frazer and E. J. Hoffman, 1912. 20 pp., 1 pl.

TECHNICAL PAPER 8. Methods of analyzing coal and coke, by F. M. Stanton and A. C. Fieldner. 1912. 21 pp.

TECHNICAL PAPER 9. The status of the gas producer and of the internal-combustion engine in the utilization of fuels, by R. H. Fernald. 1912. 42 pp.

TECHNICAL PAPER 10. Liquefied products of natural gas, their properties and uses, by I. C. Allen and G. A. Burrell. 1912. 23 pp.

PUBLICATIONS OBTAINABLE FROM THE SUPERINTENDENT OF DOCUMENTS.

The following technologic publications of the United States Geological Survey can be obtained by sending the price, in cash, to the Superintendent of Documents, Government Printing Office, Washington, D. C.:

PROFESSIONAL PAPER 48. Report on the operations of the coal-testing plant of the United States Geological Survey at the Louisiana Purchase Exposition, St. Louis, Mo., 1904; E. W. Parker, J. A. Holmes, M. R. Campbell, committee in charge. 1906. In three parts. 1492 pp., 13 pls. $1.50.

BULLETIN 261. Preliminary report on the operations of the coal-testing plant of the United States Geological Survey at the Louisiana Purchase Exposition, in St. Louis, Mo., 1904; E. W. Parker, J. A. Holmes, M. R. Campbell, committee in charge. 1905. 172 pp. 10 cents.

BULLETIN 290. Preliminary report on the operations of the fuel-testing plant of the United States Geological Survey at St. Louis, Mo., 1905, by J. A. Holmes. 1906. 240 pp. 20 cents.

BULLETIN 325. A study of 400 steaming tests made at the fuel-testing plant, St. Louis, Mo., 1904, 1905, and 1906, by L. P. Breckenridge. 1907. 196 pp. 20 cents.

BULLETIN 332. Report of the United States fuel-testing plant at St. Louis, Mo., January 1, 1906, to June 30, 1907; J. A. Holmes, in charge. 1908. 299 pp. 25 cents.

BULLETIN 336. Washing and coking tests of coal and cupola tests of coke, by Richard Moldenke, A. W. Belden, and G. R. Delamater. 1908. 76 pp. 10 cents.

BULLETIN 362. Mine sampling and chemical analyses of coals tested at the United States fuel-testing plant, Norfolk, Va., in 1907, by J. S. Burrows. 1908. 23 pp. 5 cents.

BULLETIN 368. Washing and coking tests of coal at Denver, Colo., by A. W. Belden, G. R. Delamater, and J. W. Groves. 1909. 54 pp., 2 pls. 10 cents.

O

CPSIA information can be obtained
at www.ICGtesting.com
Printed in the USA
LVHW011821061118
596180LV00013BA/1044/P

9 781528 297776